My Life and Thoughts

Amazing Facts and Very Interesting Stories including a Sci-Fi Story and Photos

by J.B. (Bucky) Maynard

Musician, Teacher, Successful Businessman, Eagle Scout, Water Safety instructor, National Guardsman, Father, Husband, and "airchair rider" still in his 70's.

Copyright © 2016

J. B. (Bucky) Maynard

All Rights Reserved.

No Part of this book may be reproduced or transmitted in any form or by any means, electrical or mechanical, including photocopying and recording, or by any information storage or retrieval systems except as may be expressly permitted in writing from the author.

ISBN: 978-0-69274367-6

Printed in the USA

Contents

I. Facts and Figures 5

II. My Life (interesting) 14

    a. My Moms Article 15

    b. My Dad's letter 28

    c. Svetlana letter 44

    d. My Philippine article 50

III. Photos 58

IV. My Sci-Fi Article 67

Dedication

I dedicate this to my family – my 100 plus year old mom
– Ann Brown Taylor Maynard,
my wife Irene, my son Lee, grandson Nic,
granddaughter Peighton, stepson Jeric,
my "birthday eating group", Dave Anders, Hames Ware,
Phil McBee
and all my friends and family.

FACTS & FIGURES

Igor Stravinsky – the music composer – said "the most terrifying thing in the world is facing a blank sheet of paper".

In movies have you noticed that when an actor opens a pill bottle he nor she never takes one pill but instead "drinks" several from the bottle.

If you feel depressed, watch a Woody Allen movie.

Why do cowboys wear long sleeve shirts especially in hot weather?

In movies, actors usually drink liquor straight without coughing.

Why is it in TV shows and movies that the person parking a car always pulls the parking brake even on level land.

What is the difference in salsa and picante sauce??

Horsepower is equal to 746 watts.

A shot of liquor is one ounce. A jigger is 1.5 ounces.

If two cars are side by side going down the highway it seems that one always pulls ahead of the other in real life.

No matter how small a project starts out it seems always to wind up major and often times you have to call for help.

A jet plane is really only a modified propeller prop plane with a gas turbine engine with the blades moved into a housing with more blades and closer together and of course a few more changes, it appears to me

Why does a bar of soap always wear in the middle??

In some western movies you can hear the wind blowing but no bushes nor leaves are moving.

Why does "looking at a light" help you to sneeze? And why does pressing the bridge of your nose help you to not sneeze.

Plants have electricity and some can feel, smell, hear, eat bugs, and produce chemical pheromones (sex chemical) plus nicotine. Bladder worts are the fastest "closing" plants at 15/1000 of a second.

The Irish are intuitive.

Who uses a butterfly net and why?

Blood is a "tissue" that makes up about $1/13^{th}$ of the total body weight.

Remember "quicksand" in the movies! Does it really exist?

Do fast foods use premium ground chuck or the cheapest ground beef (?)

Have you noticed on the Internet often are news releases about UFO's, strange objects, unusual movements and so forth. But you never hear about any of this on the news on TV or the newspapers!!!

A fortnight is two weeks.

Why is the sky blue? Answer is – it actually is not blue – it is really black but appears blue due to the refraction of light.

A subordinate clause cannot stand alone.

Why is there not a gauge in most cars to warn before the battery goes out.

Your feet smell due to butyric acid.

Why does a person yawn as soon as another person yawns? Answer is – going back to ancient times – "IT KEEPS THE GROUP ALERT".

Number of blood groups in cats is 6; dogs 5; and humans 4.

It seems your back itches at a place neither hand will reach.

Does the sun rise in the East and set in the West in the Philippines, also?

Why is the sea salty?

Why does a turkey gobble throughout the day?

Alcohols are hydroxides of organic radicals

Do you remember when an elevator man would run the elevator for you??

Lightning bugs emit "cold" light by a chemical reaction – what is it? Answer – it is called bioluminescence. They are looking for mates and there are different types of flashes.

All endings are beginnings

Luck comes in many forms.

Lead melts at 800° F.

Why are bugs attracted to light?

Aspirin is made from willow bark.

Have you noticed in movies that the actors usually pull money out of their billfolds without looking at how much money they are actually giving away such as to taxi drivers, clerks, a bartender etc. How do all the drunks in the bars in movies get home without a DWI??

What kind of whiskey has no "e" in it. Answer: Scotch Whisky! Sloe gin is not a gin but a liqueur made from sloe berries (blackthorn bush).

Microscopic yeasts consume sugar (as in corn) and the by-product is alcohol. At 172 degrees F. alcohol vaporizes.

Most people think their kids are the prettiest and smartest, that their cell phone plans are the best, and their new vehicles are the best. This also applies to over the counter meds that people take – it is hard to get them to change. Also many people will NOT ADMIT to compulsive and obsessive behavior.

Acids in many common foods and meds are – spinach-oxalic, vit.B9-folic, tea-tannic, your stomach-hydrochloric, vinegar-acetic, some drain cleaners-sulfuric.

What is the rest of the rule of "i before e except after c"?? Answer – In which case write ei as in ceiling.

Isn't it interesting that some speedometers go to 140, 160 and more when the speed limit is 55 or 70.

Colors – blue is creative, orange makes you hungry, red is sexy, certain shades of pink calm you like "drunk tank pink". This is biological in origin.

I noticed that the colors of the rainbow, in order are ROY G SIV, that is red, orange, yellow, green, blue, indigo, violet. Most parrots and some tropical birds have this same roy g biv pattern – how do you think this came to be??

HAVE YOU EVER BEEN SLAPPED? Wasn't it a shock?

My dad had 7 boys and no girls in his family. The laws of chance for this is 1/128.

I can remember when the light dimmer switch of an auto was on the floorboard.

On May 18, 1910 the earth passed through the tail of Hailey's Comet.

Maybe someday water barges will have jet motors so they will not have to move so slowly.

Catch up on your reading while waiting in the Drs. Office, bank and grocery lines, etc.

Why didn't an adventurous cowboy ever get a thoroughbred, say from Saudi Arabia, so he could outrun most of the other horses??

Also why didn't some entrepreneur teach karate to a cowboy.

Unripe pecans smell like merthiolate.

A millipede smells like iodine.

A marmoset always gives birth to twins.

Peanuts are not really nuts. They're "legumes"

Remember in grade school how after sharpening a pencil, we would blow the tip to get wood shavings off.

One benefit for a man of getting old; we save money on haircuts.

Remember when gas stations had someone to pump the gas for you.

What runs but never walks, has a mouth but never talks, has a bed but never sleeps? Answer: river

A man was asked his secret to a successful 50 years of marriage. His answer was "I keep my mouth shut"!

What word do you break when you say it?   Answer: silence

Fire is actually a "liquid", and partial plasma. It forms a gas but is not a gas or a solid.

I have never met an old person named Amber, Danielle, Tiffany, Sonja, Sasha, Jasmine, or Angel.

Vacations are not 100% fun – only 30%. Deduct 20% for expensive, 20% waiting, 10% tired, 10% chance of crashing and 10% inconvenience.

An object feels cold, not because it is but because it draws heat from your hand. Actually, in a room everything is the same temperature.

What is the purpose of a raccoon's black eyes and tail rings?

Someday instead of a garbage truck we will have a suction tube in our house to suck the garbage to a distant location.

Why do people always duck down when leaving a helicopter since the rotor blades would never hit them.

Amphibians are modified fish. Reptiles are modified amphibians

An onion makes you cry due to sulphur. Cinnamon is the aromatic bark of the laurel tree. Garlic is a member of the lily family, along with asparagus, leeks, hyacinths, onions and tulips. Raspberry is of the Rose family along with the blackberry. Tequila comes from the century plant. Sweet potatoes, carrots, turnips, and beets are "roots". While Irish potatoes and asparagus are "stems". Spinach and lettuce are leaves. Cabbage is a "bud". Celery is a "leaf stalk". Cauliflower and broccoli are "flowers". Corn, rice, wheat and rye are "fruits". Squash and tomatoes, are "fruits"

Sweet Potatoes and Yams are not the same. True Yams are rarely commercially grown in the U.S. Yams probably originated in Africa while sweet potatoes were cultivated in Peru. Yams grow on a tropical vine and can grow up to five feet in length.

The "Snake Plant" – also known as the mother-in-law tongue has sunlight type patterns sort of like a polygraph.

Anti-venom is made from horses.

Methane hydrate burns and is maybe responsible for the Bermuda Triangle mysteries.

A wasp sting is methanol. A skunk spray is oily sulfur and alcohol but it smells like transmission fluid to me. Hydrogen sulfide is what gives rotting eggs their smell.

Insects do not see yellow light. A euglena is both plant and animal.

Watts = volts x amps. Amps is a current like horsepower. Voltage is like gasoline and is a pressure.

You can make fire out of ice.

Glass is not a solid.

You can season meat with gunpowder.

Horsepower is 330 pounds raised 100 feet in one minute.

For snakes remember – "red to black venom lack. Red to yellow kill a fellow."

If you could become invisible – what is the first thing you would like to do???

Why cannot some pilot drop a bomb into a tornado or hurricane to bust it up??

Do movie stars watch movies?

Why don't more men wear flip flop type sandals without socks like women do, especially to work!

Lines that slope to the right make people feel more comfortable, so arrange your clothes so they rise to the right in the closet.

Why do we not see many 50 cent pieces nor $50 bills?? What ever happened to the $500 dollar bill? I wonder what date the world will have its first trillionaire.

A quadrillion is a number that requires 15 zeros. It is equal to 1,000 trillion.

Like the sun, tornados move north during the spring.

Why doesn't someone invent a "reverse microwave" to cool something very quickly like a Coca-Cola, etc,.

MY INTERESTING LIFE

My middle name – Buchanan – came from Donegal County, Northern Ireland – King's Calvary. Another of my ancestors, Capt. Maynard – stopped Blackbeard the Pirate. My grandmother said another relative – Ken Maynard – was the first singing cowboy in the movies.

Another relative – William Stith – was very influential in the founding of William and Mary College. My grandfather William Wallace Taylor was named after the famed William Wallace of Scotland.

My dad (Joe) and mom (Ann) are from Pine Bluff, Arkansas where the longest bayou in the World starts (Bartholomew).

MY dad was a warrant officer on a merchant marine ship in WWII. The ship name was William S. Kendrick – Liberty Ship. He was a telegraph operator officer and said at times he would have to remember a line of letters and numbers as long as a box car. He said they were always afraid of getting hit by a torpedo. My mother waited in Boston and had to share a bathroom and kitchen with two other families.

My Mom's Article

Mrs. Ann Maynard

(Lodge #605)

On December 7, 1941 when we heard the news about Pearl Harbor I knew my husband would want to get into some branch of the service. Sure enough, he joined the Merchant Marines.

While he was training at Boston, Mass. on Gallups Island, Joe, myself and two children, shared a kitchen and bath with two other families and lived in the back bay section of Boston on the third story of an old big house.

When Joe knew he was going to go overseas he come back with me and the children by train to Pine Bluff. I was on a different section of the train to get the children to sleep. Joe tried to find us the next morning, but they had switched trains in the night and we couldn't find each other till the next day in St. Louis, Missouri.

The children and I stayed in Pine Bluff until Joe could come home. When we heard the good news that the war was over we knew we would all be together again.

***

When I was in the Boy Scouts we went on a canoe trip in Minnesota and Canada in 1957. There were over 16 portages at which we had to carry the canoe on our shoulders. I wore out my shoes and had to tape them together. Our cook was drunk when he packed the food and we ran out after one week of two. Luckily there were a lot of blueberries on the islands and we ate them in as many ways as we could. We found a bag of pancake dough on the trail so we ate a lot of blueberry pancakes. The water was so pure you could drink it right out of the lake. I encourage all young men to join the Boy Scouts. I was an Eagle Scout and have used many of the things I learned while earning the Merit Badges such as swimming, camping, etc. I hate to see the Boy Scouts dying out – it is such a great organization.

In the 1950's my father had Pine Bluff Riding Stables here in Pine Bluff, Ark. He would rent horses to ride for $5 a day. I made my first $75 selling sodas at polo matches. I remember that he would play music in the stables and the horses would stick their heads out of the stalls to hear the music!!

One pastime I had during high school was to go to the ROB Roy train bridge over the Arkansas River. It is the 3rd longest U.S. River. It was not for pedestrians but me and my friends would walk the tracks to the middle and climb up to a room high up where the gears were. Several times walking back a train would come roaring around the bend and we would have to decide whether to outrun it or hang off the sides. One time I was hanging and hot steamy water out of the train barely missed me.

Another fun experience was when my friend Scott McGeorge let air out of his Chevy tires and drove us down the railroad tracks. I never realized that a car would fit exactly on the tracks. Scott would laugh when I would say "I can't be killed". And at this time in my life I really believed I couldn't die!!!

Among other things I enjoyed was attending Boys State and playing my tenor saxophone in dance bands. I started a rock band in high school in the fifties. I continued playing through college. At Vanderbilt in Nashville, Tenn. I played with the Lancers and we played a lot of fraternity dances. The writer of The Rocking Pneumonia and the Boogie Woogie Blues asked me to travel other states during Christmas but I was homesick and regret to this day that I did not go.

One event we played at was the all male university at Sewanee. Tenn. They had 2 to 4 parties a year so you can imagine how wild they were. One room was full of mattresses! I was playing a rift on my sax and a 200 plus pound drunk football player got a laugh out of pouring beer down the bell of my sax.

I got offers to play at clubs nightly in Little Rock, ARK and seriously considered it.

I still have my Selmer Paris tenor and play when I can. *(see newspaper clippings/photos next page)*

Delegates to Boys State

These eight Pine Bluff boys left yesterday for Camp Robinson and the annual session of Boys State. Left to right, they are, front row, Benny Welch, son of Mr. and Mrs. G. A. Welch; Bucky Maynard, son of Mr. and Mrs. Joe Maynard; Rod Chastant, son of Mr. and Mrs. Waldo J. Chastant; and Roy Murtishaw, son of Mr. and Mrs. Lamoine Murtishaw.

Back row: Johnny Owen, son of Mr. and Mrs. L. L. Owen; Richard Coleman, son of Mr. and Mrs. Hugo Coleman; Wendell Wood, son of Mr. and Mrs. R. W. Wood; and Bill Brudner, son of Mr. and Mrs. Edward Brudner.

ARTICLE: Four PBHS junior boys were recently chosen to be delegates to the American Legion Boys State at Camp Robinson in North Little Rock, May 28 to June 4, according to Mr. George Keely, who has charge of Pine Bluff's Boys State delegates.

These delegates chosen and their sponsors are: Bucky Maynard, Rotary Club; Roy Murtishaw, Kiwanis Club; Rod Chastant, Lions Club; and Bill Brudner, the American Legion.

Only those boys with outstanding qualities in leadership, character, scholarship, and service are chosen to go to Boys State.

A candidate to Boys State should have the following prerequisites; have completed his third year in high school, be mentally alert and physically clean, be enthusiastic, honest and thrifty, have already established himself as a potential leader, and be able to get along with others.

Boys state is an eight day training course provided by the American Legion, in study and practical application of problems of self-government. It is planned so "Boys Staters" may put into practice the theories of government as taught in the classroom, through organized political parties party caucuses, party conventions, party primaries, and general elections. The elected officials serve in city, county, and state offices.

'Journey To Center of Mirth' Holds Talent, Enjoyment For All

apx. 1960

Bel-Airs

Boom! goes the drum. And the Bel-Airs are at it again whether it be at practice or on the stage of Pine Bluff High School or at a college or high school dance. The band consists of a drummer, two saxes, two guitars, a trombone, a bass and a piano, giving a grand total of eight members in the band.

The band started out with five original members—Happy Caldwell playing drums, Bucky Maynard, and Tommy Bellhouse on the saxophones, Jackie Hendrix on piano, and Bill Nix on the bass, and acting as business manager. Their common interest for forming the band was the simple fact that they all liked music and besides it was a good way to pass off time and to work off extra energy. Then they received an offer to play for a dance and have been doing so ever since. They have been playing for a little over a year now and most of the practice sessions are held in "Happy's Doghouse."

Gradually three more members entered the band. Dewey Wallace and Don Massey playing guitar and Bryan Eans playing trombone. Occasionally George Lemmons fills in on drums for Happy. All the boys sing a little but Don Massey remains their feature singer.

Bongo Boys!

These three boys, Rick Bell, Tommy Bellhouse, and Bucky Maynard, have suddenly become avid fans of bongo music as will be shown in Saturday night's presentation of the annual Talent Show.

New Year's Eve Frolic at YWCA For Y-Teens

Ringing in 1960—with the assistance of a fire bell donated by the Pine Bluff Fire Department — will be about 200 members of the YWCA Junior Y-Teens and their dates at a semi-formal New Year's Eve dance tonight at the YWCA.

Covering the ceiling of the gymnasium will be hundreds of multi-colored balloons and crepe paper streamers.

In the center of the refreshment table will be a huge cake decorated with bells. Suspended above the cake will be a baby doll representing 1960.

Music for dancing will be provided by the Bellaire orchestra.

Serving on the adult committee which has charge of arrangements are Mrs. J. C. Edwards, Mrs. Hunter Stuart, Mrs. Don Owen, Mrs. Tommy Whiteaker, Mrs. H. G. Rogers and Mrs. Ray Mitchum.

Storybook Ball For Jr. Cotillion Saturday Night

1959

Prizes will be given for the best costumes at the annual Story Book Costume Ball for Junior Cotillion Club members on Saturday night at the Oakland Club.

Winners will include three boys and three girls in the junior high school group. Judges will be from senior high school.

Hours of the Ball will be from 8-11 p.m., with dancing to the music of the Bel Aires.

Mrs. G. N. May, Club sponsor, will be assisted by Mr. and Mrs. James T. McFall.

The Bel-airs — my 1st band

One funny remembrance from high school is some friends had an old car in which they could lift up the rubber over the "hump" on the floorboard and see the street. One time they were drinking beer and got stopped by the police so they lifted up the rubber and sat the beer on the street and pulled the rubber back to hide the beer. Boy did that come in handy.

I give thanks to God for this next event.

In high school I had broken up with my girlfriend and was so depressed that I put a hose up the exhaust pipe and through the window of my car but luckily came to my senses and killed the motor. I have only recently told this to only 2 people, I felt so embarrassed.

One summer in the 60's my friend Jim McBurnett and I went to Mexico. We flew on a Mexican airline named Siasa (Aero Navis de Mexico). What an experience! It was a DC3 "taildragger". The cockpit had a curtain for a door and it was raining through the cockpit windows and blowing in on us. There was no a/c so the open windows were the air system. There were also live chickens in crates in the back.

It took off and landed on the ground – no tarmac!

In 1967 I was a passenger in a car wreck-before we had seat belts. My head went through the windshield and back. I was hospitalized for 2 weeks and lucky to be alive. I found out I had a rare blood type – B negative. *(see newspaper clippings next page)*

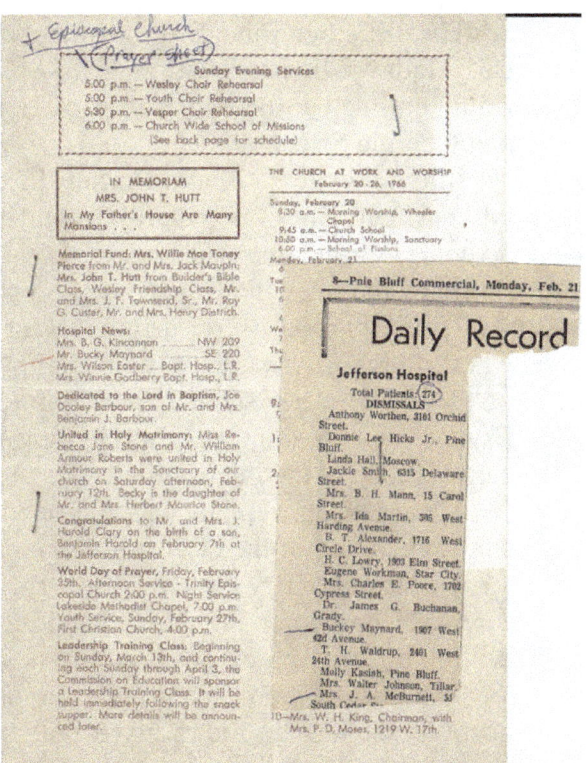

Two Youths in Car (Left) and a Couple in Car (Center) Were Injured in Collision; Parked Car (Right)

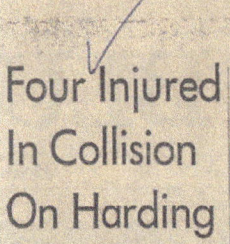

ARTICLE: Four Pine Bluff residents were injured early today when the two cars in which they were riding collided at the intersection of Harding Avenue and Georgia Street.

Injured were William Bryant Jr., 22, of 3705 Olive Street; Bucky Maynard, 22 of 1907 West 42nd Avenue; Willie Lee Jackson, 40, of 611 East 24th Avenue; and his wife, Rosey Lee Jackson, 43.

The Jacksons were driving west on Harding when they attempted to turn south on Georgia, police said. The second automobile, driven by Bryant, was traveling east on Harding when the two cars collided at the intersection.

The Jackson vehicle was pushed into a parked car by the force of the collision.

The front of Bryant's vehicle struck the Jackson car near the center on the right side. Jackson was driving his car. His wife was sitting next to him.

The injured were taken to Jefferson Hospital. A hospital spokesman today listed both Jackson's and his wife's conditions a fair. Jackson sustained several fractured rib, cut and bruises. His wife suffered a fractured leg and a sprained back according to the spokesman.

Here is a formula that Mrs. Watson, my high school algebra teacher, gave us to show that 2=1.

$$\text{Let } A=B$$
$$A^2=AB$$
$$A^2-B^2=ab-B^2$$
$$(A+B)(A-B)=B(A-B)$$
$$A+B=B$$
$$2B=B$$
$$2=1$$

Mr. McGinnis – my chemistry teacher at summer school at UofA Monticello said he never had a head cold due to sniffing chlorine gas each year during lab experiments. At Henderson College, where I transferred after Vanderbilt, I tried to stay in economical places to save money. One place, a one room apartment, had no curtains; only pull down shades was $30/month. The "oven" was a metal box on top of a burner. The shower was under some stairs and I had to duck down to shower. Another room I rented an extra bedroom inside a house with an old man whom would sometimes bust into the bathroom and order me off the toilet immediately as he was having a diarrhea attack. If I came in too late I would be locked out and have to crawl in through a window.

Another apartment I rented in Henderson College in Arkadelphia has a funny story with it. I had previously told a Pine Bluff friend that he was welcome to visit me there. I later moved and had not seen him for some time to tell him. His name is Steve Huselton. I later ran into him at a Homecoming game and he thanked me for letting him change clothes after a shower. What a laugh. Just think what the current renter – male or female – would have thought if he or she came in and Steve was in the shower.

I was in Theta XI fraternity and as a pledge I lived in the frat house I had a hard time studying due to the loud noises, hazing, etc. So I asked different preachers if I could study at their churches and they would often check on me to be sure I was telling the truth.

One night at the Theta XI fraternity house myself and 2 friends were studying. Two students came by and said they were going to Hot Springs – a nearby city where beer was sold. A few hours later they came back and were either on drugs or drunk or both. They were changed and different. They made us stand at attention while holding a broken glass coke bottle near our throats. We could not smile. The police station was caddy cornered across the street and I thought about making a wild dash for it but I was too scared and closely watched by the "thugs". It was sheer terror for what seemed like hours but finally they left. They said for us not to tell anyone or they would retaliate. To this day I regret not turning them in.

I majored in pre-med and talked my advisor Dr. Basford into offering Embryology at Henderson. One time I was taking Comparative Anatomy and we were studying a very huge cat. The muscles and veins were all separated and pulled apart and color-coded. Basford said I could take it home to study so I put it in my V. W. Beatle in the front where the gas tank also was. I later forgot it was there and pulled up to a gas station to fill up.

This was when the gas attendants would still put your gas in. When he raised the trunk I heard a loud yell and he went off running.

P.S I passed the medical school admission test but due to low grades my first year at Vanderbilt I was put on a waiting list at #8.

During that summer I taught Life Saving, Swimming, and Diving at Eden Park which had a million gallon pool and is said to be one of the largest private pools in the world. One day I had to rescue a lady trying to swim to the deep part and it ruined my watch. The next day the same thing happened with another lady and I ruined a second watch! During that summer I moved to #1 on the med school list and was advised to work on a Master's degree. Carter Short, my relative and the Registrar got me admitted at the last minute to a U of A grad school. I finished the semester but lost interest and ran short of money.

Also even though I made A's in chemistry as my minor I did not do well in Biochemistry in grad school. I felt like a failure at a medical career, so I entered the business world.

My dad started Newcomer Greeting Service in 1952 and at one time had it in 180 cities and 28 states. I worked with this and traveled, hired field reps, hostesses, and learned sales. My dad truthfully taught me that is you can sell you can always get a job. One time I was in Monroe, LA. hiring a hostess and one night I went to a restaurant which had a lounge band. For one song they cut the lights and you could see the "glow in the dark gloves" of each musician. It was really unique. For years after this I would see the same guitar player as Jerry Lee Lewis's backup guitar player. He was good.

Attached is a letter my dad wrote to a field rep. in Alabama in 1969. As you can see he was a real sales motivator!

My Dad's Letter

December 30, 1969

Mrs. Miriam Carroll
1680 Beckham
Birmingham, Ala

"Just a few thoughts looking forward to the New Year."

I wonder if you have any books on salesmanship on hand? If no, I would like to suggest that you try the public library or pick up a volume from the newsstand. These paperback editions are worth their weight in gold. I am a firm believer in a salesman staying in condition the same as an athlete staying in condition. With these "soreheads" all selling you, telling you how broke they are, how bad the times are, why the Newcomer Greeting Service Program isn't any good, etc., you need a page or two of good sales literature at noon to give you the lift to carry you through the afternoon on a positive plane.

You are working as a salesman, which is in my opinion, the very finest profession there is these days. This is, frankly, the highest paid profession.

It is a crying shame, however, that we have this wonderful profession muddied up so by people that are not qualified or those that will not even attempt to prepare themselves. A doctor prepares by attending college for eight years and then serves possibly two year internship. This is ten long years spent singly learning how to do the job. How many years, how many months, how many weeks, or how many days have you actually worked at preparing yourself for this biggest of all professions, the professional salesman? I do not want you to answer this question because I'm afraid it would be embarrassing.

Before a doctor performs an operation or makes a diagnosis on an important case, he will review. He thinks about his campaign as he reviews and makes very careful plans. Before you make a sales call, do you review and do you plan your campaign carefully? I believe one of the biggest reasons why most salesmen fail is that they never bother to make any plans, in the slightest, or prepare themselves for this big interview that is coming. A championship prize fighter would not think of going into the ring unless he went in training. The greater the professional, the more he stays in training. What are you doing these days to stay in training? I will not ask you to answer this question either because I am afraid it would be embarrassing. It is best to plan in the evening for the next day or take time out the first thing of a morning to plan your calls for the day. I am talking in terms of not only your route but the type of businesses to contact, the individuals you are going to talk to, what the items are that you know they will be the most interested in that you want to push and the pitfalls you want to avoid, etc. one of our successful map worker these days actually lays out an ad for the merchant she plans to call on the next days, shoving him what his ad is going to look like because she means to sell him this particular space. We are not going to say this is the best program to follow but this individual is doing very well selling the big deals with this type program.

Most every successful sales person that I know of is a hard worker. We believe that success, when you add it up, amounts to about 98% perspiration and 2% inspiration. Getting in an extra hour with sales each day for a five day week means an extra half day added to the week. Some people get in two extra hours a day by working just a little later in the evening.

Most successful people that I know are "positive thinkers". Every word they say and every move they make radiates "positive thinking". They simply do not have negative thoughts. I am sure you know that negative thoughts are transferable. Positive thoughts are also transferable. I hope, during the New Year, that you are going to start off thinking positive and keep thinking positive throughout the entire year. Unless you are sure you are going to get this job done, you will not have a chance of being successful.

Finally, I believe that enthusiasm is catching. Where you find a person getting enthused over the project they are working on, you will also find this project not only is a project that gets finished but it gets finished on time and in proper order. In today's competitive markets, enthusiasm is almost a must if you hope to finish a project.

Preparing yourself for the task at hand and then going to work putting in hard work with positive thinking plus enthusiasm is an absolutely unbeatable combination. I do hope, as you make plans now for the New Year, that you are going to be putting reminders in every direction to follow a program that will call for preparing for the job, plus enough hard work with positive thinking and enthusiasm. If so, I can assure you, that 1970 is going to be a very great year for you.

Yours truly,
Newcomer Greeting Service

Joe B. Maynard, Manager
JBM;mjr

***

I still have the greeting service going as I write this. In the meantime, the Vietnam War was going full blast. Although I had had 2 years of ROTC at Henderson I applied to officer's school. When the committee learned that I had a B.S. degree in Biology they offered me a "direct commission" due to the emergency of the war. So as I waited I learned I would be a Lieutenant in the medical service corps, probably on the front lines so some friends talked me into joining the National Guard which I did for 6 years. I had a secret clearance. At the end of each summer camp we had water fights with fire hoses. I later went to the Guard OCS School at Camp Robinson.

At OCS school we only got 3 hours of sleep per night. One day we came in from climbing up a steep hill with gas masks in 100 degree weather and all our foot lockers were stacked on top of each other to above the door so we could not get in.

This was due to one cadet not making his bed right. One night we were called into formation in our underwear and had to march around the base in this attire. Another night we were paired into twos with a flashlight and a compass and terrain map in the middle of the woods and had to find our way back to camp. I loved this since I am intrigued by maps even today. During basic training at Ft. Gordon Georgia I asked the First Sargent for a weekend pass. He asked why? I said – to see Billy Graham. He replied "I like him too – here's your pass." This was in 1967. Billy Graham was great and at that time only had about 300 people in the small church. Rev. Billy Graham is still my favorite today.

At Augusta, GA I also got to go to the Master's and will never forget hearing Arnold Palmer asking which was way North!!! On another weekend pass I went to Lake Charles from Rt. Polk, LA. After leaving a coffee shop in my uniform, I looked behind me and three hoodlums who had been at the coffee shop were hot on my trail. Luckily I saw a phone booth and ducked into it and shut the door and acted like I was calling and they kept on going. Whew! What a close call – the Lord helped me out again!

I am still friends with some of my former guardsman such as Allen Searcy of Kingsland, Arkansas who does taxes.

An interesting experience happened when I dated a girl named Jessie. Her stepdad worked for Central Transformer here which my uncle, Ralph Mitchell, helped found. He would fall asleep on the couch and repeat word for word his exact conversation with a fellow employee. You could not hear the employee's words but just silence when this employee was speaking. Remember the conversation was exact with no words left out like "Okay I am going to turn this screw to the right and then we will check the gauge. Silence .... No don't do that yet wait until I check the gauge", and so forth. It is a good thing that he did not run around on his wife!!!

In the 1970's my dad and I took a look at the former Shriner's building at 308 W. 2nd in Pine Bluff, Ark. It was built as the Eagle's Club originally at the turn of the century. It was full of dust and pigeons but a beautiful building. It had a lot of oak, many stained glass windows, 2 fireplaces, potentate chairs, beautiful large stairway and on and on. We had the Newcomer Offices in it at first but then he had the idea to start a nightclub. At first it was slow taking off. But then Dave Anders came by, after having a very successful club in Tampa, Florida – the Loser's. He brought his light and sound systems and the club took off. My dad had some good ideas – he knew that if the women came then the men would follow. So he gave the women free or low priced drinks to get them coming. On Wednesdays he had "SINK OR SWIM" Night that was very popular. For $6.00 one could drink all he wanted from the "bar" brands not the "call" brands. The Alcohol Beverage Control later said clubs had to charge so much per drink. The club was called Yesterday's and was at it's heyday from the mid 70's to mid 80's. Other popular nights were "wet –t-shirt contests and numerous costume parties.

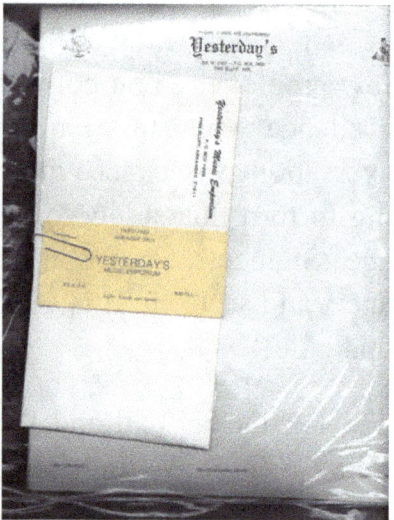

5-25-00

Patrons examine sale items during an auction Wednesday in Pine Bluff at the former Elks lodge, built in 1922. The building, later a Shriners' club and, finally a night club, will be demolished to make way for the new Donald W. Reynolds Community Services Center.

Club's end echoes groups' decline

Former lodge of Elks, Eagles, Shriners and Masons sold off in pieces

BY LINDA S. CAILLOUET
ARKANSAS DEMOCRAT-GAZETTE

PINE BLUFF — The Elk is an endangered species these days. So is the Eagle. Shriners and Masons are dying off, too, and with them all go their natural habitats.

Sold! Sold! Sold! was the refrain Wednesday in downtown Pine Bluff as the hammer came down on the former Elks/former Eagles/former Shriners/former Masons lodge.

The circa 1922 Craftsman-style building at 308 W. Second St. crossed Bruce Cooper's auction block, not in one lump sum but in pieces and parts: door by door, window by window, staircase by staircase, disco ball by disco ball.

"Come prepared to remove [most items are still attached]

The former Elks Lodge in Pine Bluff.

and haul your purchases — three days' removal time!" warned an ad in Sunday's *Arkansas Democrat-Gazette* classifieds.

Every time Cooper's hammer went down it signaled another lucky bidders' hammer going up

to nab a nail with the claw or to pry loose a hinge.

The club, originally built by the local Benevolent and Protective Order of the Elks, was most recently a night club called Yesterday's.

For the past couple decades there's been a stilln in the air on Saturday nights West Second Street downtov even around the stately Saenger Theater and its small counterpart across the street, t Community. The only signs of l came from the music pouring of Yesterday's two blocks dov But now silence pervades ev there.

The building will be razed make way for the $6.7 milli Donald W. Reynolds Commun Services Center.

Construction on the cent being built by the United Way Southeast Arkansas with a million grant from the Reynol Foundation, will begin in ear 2001. The 40,000 square-foot ce ter will house 14 not-for-profit

See **AUCTION**, Page

After the mid 80's and the Disco era the Yesterday's heyday was over. The white group went to PJ's and Yesterday's became Monroe's a predominantly black club which was also popular under the management of Melvin Ely Monroe. Yesterday's was one of the best discotheque's in the U.S.

One night I was going home and I saw a huge crowd at PJ's – a truly "swank, club on University. I was getting ready to leave and P J. (Perry Johnson) said "Bucky you can't go now." Why I asked? He said, "this is the last night of the Most Outrageous Contest and the winner gets $1000.00!" So I watched the first act – a girl with her stomach showing laid on her back and balanced a coke bottle while pushing her stomach upward by arching her back. Not too impressive. The 2nd act was started by 2 young men in suits and ties came out holding a huge live Catfish. They ground it up in a blender and each took turns chug-a-lugging it. A little better!! Then a man came out with bloodshot eyes wearing a fur coat and long hair and carrying one of the biggest Possum's I had ever seen. He cut it open, pulled the intestines out and started eating. P.J. had to send out 3 employees to mop up the blood. Needless to say this man won the Most Outrageous Contest.

My brother in law – Hubert Hankins – was a State Representative when Bill Clinton was Governor. He said Clinton would invite some of them out to the Governor's Mansion or duck hunting to play Hearts but would not let anyone leave until he was winning – sometimes 3:00 am.

In 1979 I took a sales job with Colonial Life and Accident Insurance Company. It was similar to what I had been doing with the Newcomer Service – that is calling on business owners. LEE Ray Hardy trained me and as we would enter a business in our coats and ties we would say "don't panic, we are not the IRS OR FBI!" I was successful and this advice will help many of you with sales jobs and even other situations in life. To get myself "'psyched up I would say optimistic things to myself to give me courage such as: "your dad was a good salesman and he would be proud of you; you have a B.S. degree, your uncle is a medical doctor; you went to Vanderbilt; there are attorneys in our family and so forth". I would think about my mother use to say "go do it, before you think of a reason not to."

On my first New Years Eve I called on Fast Food Managers up to 5:00pm. The next week my original trainer, Lee Ray Hardy flew in from Texas and at the airport he said "I wanted to see how the #1 salesman in the country is doing." I had no idea – I was just having fun! I later realized that all this was during deep U.S. recession!

Later the Zone Manager flew me and 2 others, to Ft. Smith for me to train a new person. We were in a small plane – just the 3 of us. I was almost asleep in the back seat and suddenly the motor quit. Dead silence. I could see and hear the cranking the motor. Finally it started and he said "I forgot to switch to the other gas tank." Whew!

I later started my own agency adding property and casualty and am training my son Lee to learn the business.

Like Donald Trump says "sheer persistence is the difference between success and failure". I never thought about "competition".

On August 23, 1992, my birthday, I was visiting my friend Jim Thompson of Miami. We flew to the Bahamas and at about 3:00 am someone knocked on our hotel door and said we had to evacuate due to hurricane Andrew (Raggedy Andy)! At the Nassau airport it was like the evacuation of Saigon – people everywhere with some running down the baggage conveyor line. Our jet could not turn off the engine due to some malfunction so we had to hold our ears to board. Back in my friends condo in Miami Jim and his wife and kids were in the closet and I was on the couch. At Andrew's peak the glass sliding doors near me were pulsating like rubber. Water was leaking through the ceiling from three stories up. The next day Jimmy said, "Oh my truck is gone"! We found it across the parking lot. Andrew was a 5 and as I looked out I could see no shingles on any roofs.

The stair ways were blocked and someone said a baby was trapped upstairs, but we later rescued it. Luckily Jimmy had a furniture warehouse and we stayed in it with the National Guard patrolling outside. There were no restaurants since no electricity to operate the cash registers. No flights out either so a week later, I rented a car and drove north to Tampa. All the cars were coming back south with very very few going my direction so I went very fast – way above the speed limit since I knew most of the cops would be in Miami. Luckily I was right.

In 2000 I went to Costa Rica. It was Super bowl weekend and my friend Jimmy bet on one team so I bet on the other team. When I went to collect my money the "mafia type" man said come back tomorrow but I said I could not. I asked why and he said he did not want anyone to see him paying me. I think he was trying to get out of paying me. I WAS persistent and he finally paid me. I thought later he could have had me bumped off especially being in Costa Rica.

I have always liked boating and water sports. On 2 occasions I went with a boating group the PIRATES OF THE MISSISSIPPI by boats from Pendleton on the Arkansas River to Greenville, Miss. on the Mississippi River. We would leave Saturday morning and after going through several locks and dams and stopping at huge sandbars, arrive in Greenville that afternoon. Some towed their boats back but we went back by boat with the smaller group. Somehow my girlfriend and I got separated from the group and as our gas got lower and lower it was getting darker. We hit a sandbar on the Mississippi so we decided to sleep in the boat having to cover with the tarp to keep off mosquitoes. We later found out that the boat uses about 3 times the gas going upstream – which we did not plan for. We flagged down every tow boat but all they had was diesel. Finally I was relieved to see a Pine Bluff Sand and Gravel tow boat. The owner of the company is still a good friend, Scott McGeorge, and he is in our "birthday group" that eats out on birthdays. Anyway one of the deckhands gave us a couple gallons of gas that got us to Rosedale, Mississippi. We had a fisherman drive us to a gas station for more gas. We got home a day late but were relieved and prayerful.

As of this day today I still like riding my "AIRCHAIR". It has a seat and a metal shaft going about 3 feet into the water with an aerodynamic fin on the bottom. It is a great feeling being "Up in the air" and not feeling the wake as a slalom ski does.

Another Arkansas River adventure is when my Labrador Hershey and I and my date pulled up to one of my favorite islands. After we walked around the island I saw my boat floating downstream in the middle of the river. I jumped in the still cold water and swam after the boat while feeling Hershey's paws on my back. She is heavy so it was hard pulling her up in the boat. I am lucky that the current did not pull the boat into Lock and Dam 4.

In 2009 I took an "Anastasia" trip to Ukraine with a bunch of guys. We had 3 socials set up to meet women. There were a lot more women than men. There were also interpreters on hand but many of the women could speak English. It was interesting and enlightening and I dated some beautiful women but never met one to marry. Most had cell phones but no autos and made very little money.

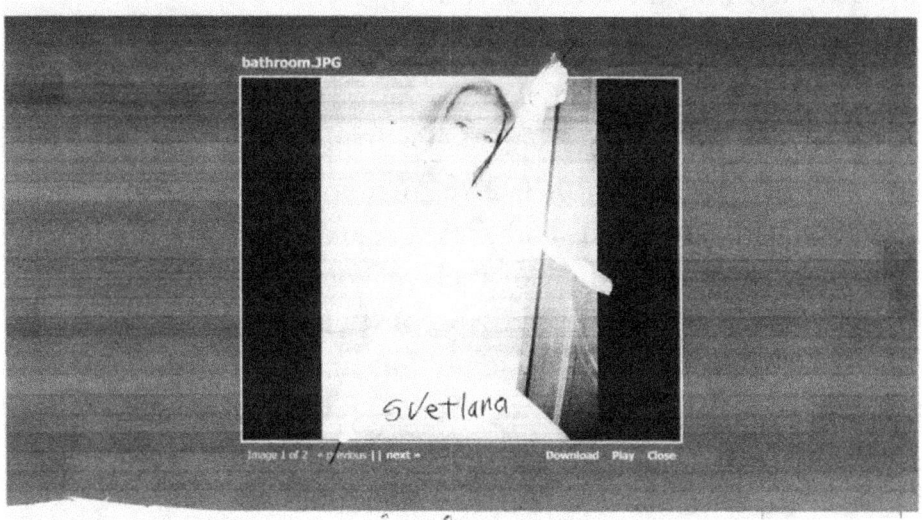

One day, in 2010, I got an email from Svetlana whom said she was from Russia. I had recently returned from an "Anastasia" trip to Ukraine so I was interested in seeing what she had to say. I made sure she was not after money and as we corresponded (see some correspondence attached), she called me late one night. (It is day there when night here). She also sent me a talking dvd. As you can see from her emails she finally started asking for money.

Svetlana Letters

FROM: svetlana (navaoba@yahoo.com)

TO: jbm…

DATE: Tue, November 30, 2010 2:00:10AM

CC:

Subject: I want to be with you

Hi My Love Bucky!

I am very glad to get your letter. I have no apartment in Moscow, I rent it. But I have own apartment to Pskov. I work with documents and files, I also am engaged in taxes of our hospital. I did not receive yours airchair photo. I understand you. I will tell to you if I do not receive the visa. But we should wait some time and I will tell to you I will have the visa or not. Today I have visited airport Sheremetyevo, and have considered the prices for air tickets. It was necessary for me to find cheap air tickets. The employee of the airport has helped me at a choice. I have considered air tickets for December, 23 to Little Rock National Airport. Cost of one air ticket makes 720 us dollars. It is necessary for me to have 1440 us dollars that I could get air tickets. Bucky, already I said to you, that your help will be necessary for m. now I do not have money to get air tickets. I have paid registration of the visa and the medical insurance in embassy. Also I have paid residing at a room. At me now remains 150 us dollars, but these are necessary money for me to live in Moscow. You can help to get to me air tickets? Already I said to you, that I should get air tickets in Russia. Today I also have visited bank. It was necessary for me to find out, how I can receive from you money. Never earlier I faced it. In bank to me have explained, that in the world there are many systems for remittance. We can use Western Union that you have transferred money to Russia, and I have got air tickets. Now I will tell to you the information which is necessary for you that you could transfer the help to me.

BINBANK
Grodnenskaya street, 5a
Moscow, 121471

My full name Svetlana Sobyanina.

After you will transfer me the help, the employee of bank will inform you MTCN. You should tell to me MTCN that I could receive your help. I hope that you have understood me. When I can receive your help? The earlier you will help me, the my visa will be faster ready and we can meet. Please write me as soon as possible. I will wait for your letter. I miss you! Only yours, Svetlana.

***

Moscow. Today to me have informed the important

information ... since recent time at the airport have entered life insurance system. I should buy an insurance policy. The insurance costs 18000 rubles, that is 600 US dollars and my life will be insured for the sum 28000 US dollars. The insurance is voluntary. However, I am frightened by flights by the plane. Especially I should fly on the big distance. I know, that in air crashes many people perish. Bucky, I am afraid for my life and I will necessarily get insurance. Also now in the world the difficult situation, some airline detain the flights. In case of insurance, I can receive back money from acquisition of air tickets in the event that the start will be canceled. The last what I will require for arrival to you, this acquisition of air tickets. Now I do not know how many air tickets will cost ... I will know it when I will be in Moscow. Now I have some savings. I have 36000 rubles, that is enough this money that I have issued for the visa, and live in Moscow, and also has bought the insurance. Bucky, at me does not remain money to get air tickets. You can help to get to me air tickets? It is important, and I wish to pay the big attention to this question. I cannot pay all expenses... I do not have enough of money it is not a shame to me to speak to you about it. I trust you and consequently I ask you the help. Now I do not know, will cost how many air tickets, it is necessary for me to visit the airport and to find out about their cost. I plan to arrive to Moscow with a few next days and then I will speak you exact cost of tickets. I think, that division of expenses it is fair. It is necessary for me to be assured that you wait for me, and you will meet at the airport. I think, that cost of air tickets is not big, and you can help our happiness. Soon we with you will together. I have already prepared the report for my director. Likely already tomorrow or the day after tomorrow I can have a vacation, and I am ready to go to Moscow to visit embassy and to begin process of registration of the visa. Now I have some questions for you now. Bucky, what things I should take with myself to feel comfortable in yours country. I with impatience wait for your letter. Please, write to me and tell, what you think of the information which to me managed to be found out today. Bucky, <u>I love you!</u> I promise, that I will be the most beautiful woman at the airport and you will be proud of me. I found some of my old pics. I sent it specially for you. Nobody see this pics. With kisses, yours Svetlana.

***

Wednesday, December 1, 2010   5:50 AM

From: "Svetlana" navaoba@yahoo.com

To: "Bucky Maynard"

Hi my second half Bucky!

How are you today? I hope you fine, I am okay, but I will better when we will be together. I very much miss you. I think of you all the day long. You always in my heart. I am very glad that you will help me. It is very fine. It means that very soon we will together. I am very excited by our meeting. I can arrive to you on December 17th. I will be at you month and I leave on January, 17th. If it suits you? Or you want that I was at you other time? I should buy tickets in Moscow independently. The embassy does not accept a copy and a photo of tickets because it can be a fake. So I was explained by the consul. I am obliged to buy tickets itself. I heard that you have told that Love me. Yes, I am excited. I love you too! I cannot name to you number the visa business because the consul has forbidden me. I signed agreements on which I cannot advertise data. It is connected with struggle against terrorism. I understand, that you not the terrorist, but I do not wish to have a problem with it. I hope, you understand me. I did not speak you the name of my hospital because you did not ask me about it. I work in the Pskov Regional Hospital. I have no phone in apartment which now I rent. But I have your phone number and I will call to you from post office of telegraph tomorrow. Please hold your phone nearby. I wish to hear your voice. Navaoba it is simple a word, a set of different letters it does not mean anything. I love you. I wish to hold your hand and to walk with you. I wish to fall asleep with you. I wish to wake up from your gentle kiss. I am very glad, that soon all our dreams will a reality and we will be together. I am grateful to you for your love I the most successful woman because I have found you. You my man and I will not give you to anybody. You mine! I want you! Please, write to me as soon as possible. I will wait your letter. With love, Svetlana

***

I had her send a copy of her passport and I contacted Senator Mark Pryor's office and they said it was a "FAKE"! She was on the Anti-scam list.

Another harrowing experience was with a girl I dated named Shannon. For whatever reason – maybe an overdose of her prescription meds – she started having hallucinations. She was swatting what she believed to be spiders everywhere. I got her into my vehicle and as she was still swatting and trying to climb out the sunroof I had to run every red-light to the ER. It took 3 orderlies to get her out of the vehicle

Interestingly enough, after traveling to Ukraine to meet a woman, I got a call on New Year's Day 2011 from a friend's girlfriend who said she had a Filipino friend in San Francisco who was single. Her name is Irene and one thing led to the next and we later got married. Irene is great caring Christian and everyone loves her. I thank the Lord for bringing us together.

MY INTERESTING TRIP TO THE PHILIPPINES IN JANUARY 2014

My wife and I recently returned from an amazing trip to the Philippines It was my 15th time there and her first time in nine year. We left home about 3:30am New Year's Day for Little Rock, Ar., and passed a few night clubs still going strong. The flight from Dallas to Seoul, Korea was 15 hours. I can't sleep well on a plane, I guess I'm paranoid that it will crash, but the service was great on Korean Airlines and beautiful stewardesses.

The flight to Manila was another 3-1/2 hours. Manila Airport was hectic, to say the least, later at night. I was relieved to see many policemen at the airport. We took a van to

To my wife's home town area of Balatan It turned out to be a 12 hour drive, I strongly suggest the bus over the vans, including the longer "Jeep" vans. It was nerve wracking most of the time. 80-mph (stop-n-go). No seatbelts, packed in like sardines but our driver Marlon loves to drive and got us there safely after many harrowing experiences.

Balatan is probably typical of many Philippine areas, very loving but impoverished people. No running water, few refrigerators, open doors and windows, cooking in big pits outdoors, straw roofs, hand pump well. My wife's family was a few steps away from the ocean, but spared recently from the tremendous typhoon (Yolanda). Many days we ate fish freshly caught from a wooden boat.

Last night we went to an annual festival (Saint Anthony) De Padua for the area and were honored by a very nice chief of police, and several aldermen.

***

Today many men put up a tent in honor of my wife's mother's one year death.

Tomorrow many people will bring food after church. The alderman donated a nice wood table, etc,. My wife's relatives fixed pancit, a local dish, for the workers.

These are a loving, constantly sharing, impoverished people whom share hand-me-down traditions for decades.

Being married to a Filipino I have learned that any time friends get together a Philippine dish like *lumpia, pancit, nilaga*, etc., is cooked by both sides.

The women are beautiful, the men hard-working but both caring and loveable, and thoughtful. Their houses are so close that there are practically no yards.

In Balatan, there are almost no vehicles but a lot of motor cycles (tricycles with cages).

Priest Rev. Father Diosoro Ibanita performed a mass for my wife' mother's first anniversary of her death. I couldn't understand the Togalog but afterwards he come to bless the house in English. I'm not sure if it was for me or traditional but at least I could understand. He reaffirmed what a generous, close and caring the Philippine people are, and that the Philippines includes over 7,000 islands. I told him how I used to hear Mike Huckabee preach in my hometown of Pine Bluff, Arkansas. He had heard of him.

One of my most enjoyable past-times was saying "Kuma sta" to the kids as they walked down the dirt road in their school uniforms. Many seemed shy and most of their eyes would follow me as I walked or as they walked to schools.

Most would say "bye" with a big smile as if they'd never seen an American. A neighbor said I was like an "alien" to them. Their holidays and festivals are expanded to many more days than ours.

We may wait three or four years before returning due to the expense and long travel hours, and then I hope to write another experience.

Bucky Maynard

PS: In Manila – traffic signals are "rough guidelines".

***

I have always loved music and have been fortunate to see in person: Elvis (twice) Santana (this was at the annual New Orleans jazz fest in April) and I asked a man behind me if I could buy one of his cokes since it was a huge crowd but he said no. Later after he left I saw he left a full unopened can at my feet. Also I have seen, Jerry Lee Lewis (he flipped a half full bottle of beer over and over and each time it spewed the front row people), B.B. King and Bobby Blue Bland (front row center – a huge lady next to me was rockin and almost knocked me out of my seat), Chicago twice, Tom Jones, Little Big Town, Elton John and Billy Joel, James Brown, Mike Huckabee's band (several times), KISS (my whole body vibrated). Also ZZ TOP, Steve Martin (banjo), Hank Williams Jr., Kool and the Gang, Beach Boys, Earth Wind and Fire, Dionne Warwick, Journey (with Filipino singer Arnel). Stevie Wonder and more. I recently saw the Eagles with my wife, Irene. Can you believe they are still playing now in 2015 after starting in the early seventies!!! Isn't their harmony fantastic!!!

Rock 'n' Roll Singers To Appear

The Pine Bluff Junior Chamber of Commerce will present three rock 'n' roll groups in a concert here December 26. The concert is the first of a series that is intended to include various types of entertainment.

Jerry Lee Lewis, Ace Cannon and The Tarantulas are scheduled to appear at McFadden Field House at 7:30 p.m.

The date for this concert was set in hopes that the greatest number of young people would be out of school and able to attend at that time, according to Bill Lewis, publicity chairman for the project. Well - known dance bands may be scheduled for other concerts later, he said.

The series, which will include three or four concerts a year, is not planned as a moneymaking project, Lewis said, but as a means of providing varied live entertainment for this area.

Advance tickets for the concert will be $1 for general admission and $1.25 for reserved seats. Tickets at the door will be $1.25 and $1.50. Tickets may be purchased from any member of the Jaycees or the Pine Bluff Hi-Y Club.

JERRY LEE LEWIS

Concert Program Features Singer

ARTICLE: Jerry Lee Lewis will be featured performer in the concert to be presented Thursday at McFadden Field House at 7:30 p.m.

Also on the program are Ace Cannon with his saxophone and The Tarantulas, a five-piece band.

Advance tickets are $1, general admission and $1.25, reserved seats. Tickets at the door will be $1.25 and $1.50

The concert is sponsored by the Junior Chamber of Commerce.

Audience Reacts Stoically to Performers

This was the typical reaction last night to Jerry Lee Lewis and The Tarantulas, who performed at a rock 'n' roll show here.

Jerry Lee Lewis's Cool Show Leaves Youthful Audience Cold

The most recent musicians we saw were Barry Manilow, and the Doobie Brothers. ***

I have a great family. My former grand-daughter Sidneigh got killed in an auto accident a few years ago.

This was my grandaughter Sidneigh, she died in an auto accident.

My dad, Joe Maynard, died in 1992, but he is said to be one of the best duplicate bridge players in the state. He once played with Omar Shariff!!

Thanks to Phil McBee for his help. Also I want to thank Dave Anders for encouraging me to write this. He has written several good books you should buy, *No Title Fits* and *The Loser's – Tampa*.

Isn't it amazing how you will suddenly think of something and say "why didn't I think of that before".

I take 2 flu shots every year! I figure if one is 80% effective then the 2nd gets me to maybe 90%!

I also have learned that with some services like TV, internet, telephone service, and some others, if you ask for discounts then you will often get them!

Retired people usually "are" busier than before since they are doing all the things they previously put off.

J. B. (Bucky) Maynard

Photos

Mr & Mrs G.E. Maynard, Joe Maynard & 6 brothers

PB Physician Makes Plans To 'Hang Up Stethoscope'

Dr. Ross Maynard Reflects on His Career in Medicine

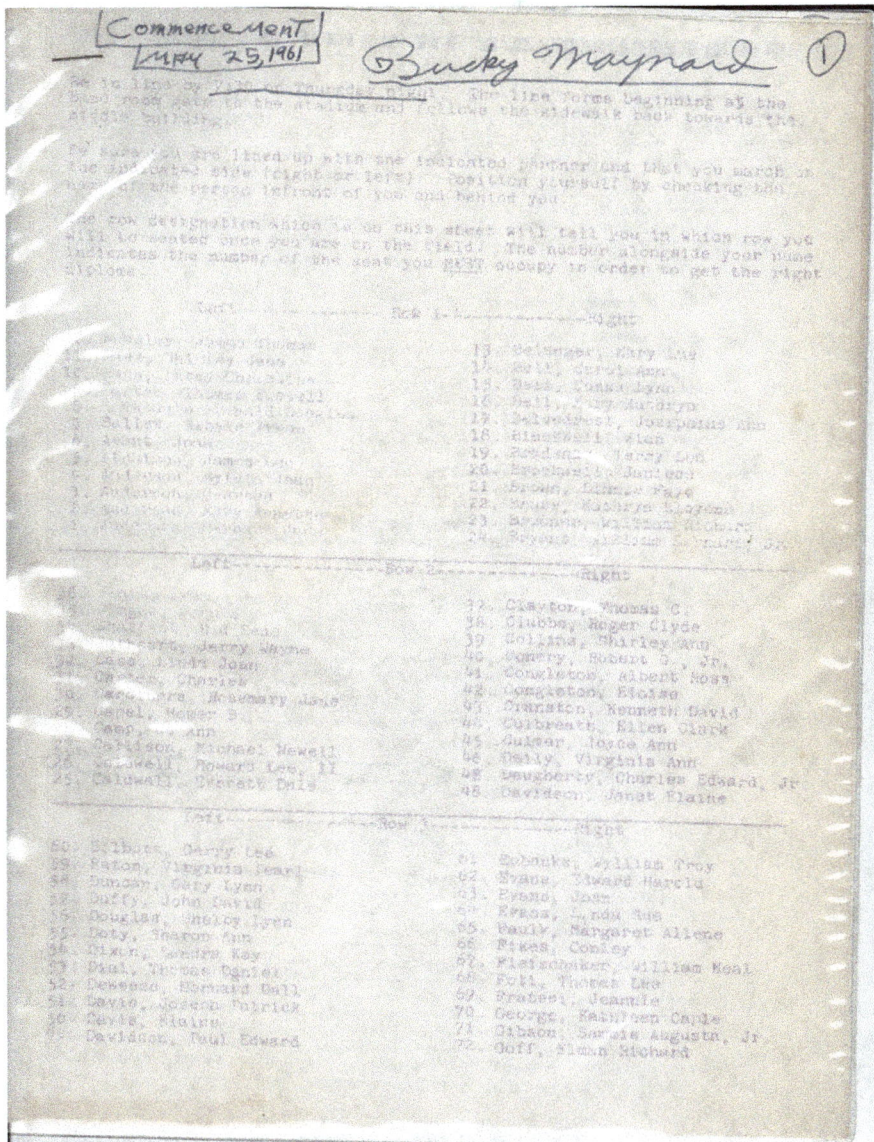

United States Senate
COMMITTEE ON FOREIGN RELATIONS

J. W. FULBRIGHT, ARK., CHAIRMAN
JOHN SPARKMAN, ALA.   ALEXANDER WILEY, WIS.
HUBERT H. HUMPHREY, MINN.   BOURKE B. HICKENLOOPER, IOWA
MIKE MANSFIELD, MONT.   GEORGE D. AIKEN, VT.
WAYNE MORSE, OREG.   HOMER E. CAPEHART, IND.
RUSSELL B. LONG, LA.   FRANK CARLSON, KANS.
ALBERT GORE, TENN.   JOHN J. WILLIAMS, DEL.
FRANK J. LAUSCHE, OHIO
STUART SYMINGTON, MO.
THOMAS J. DODD, CONN.

CARL MARCY, CHIEF OF STAFF
DARRELL ST. CLAIRE, CLERK

June 30, 1961

Mr. Hoseph B. Maynard, Jr.
2006 Laurel
Pine Bluff, Arkansas

Dear Mr. Maynard:

    I wish to extend my sincere congratulations on your graduation from high school.

    Your diploma represents a good deal of hard work and perseverance on your part and its award is an accomplishment of which you may be proud. It does, however, place upon you a greater responsibility in the affairs of your community, State and Nation. In the difficult times that lie ahead for our country, knowledge will be our most valuable weapon in the defense of our freedom. I know that you will continue to learn, and work to bring about a better Arkansas and America.

    I wish you happiness, and success in your future endeavors. If there is ever any way I can serve you, I hope you will not hesitate to call upon me.

    With best wishes, I am

              Sincerely yours,

              J. W. Fulbright

JWF:flt

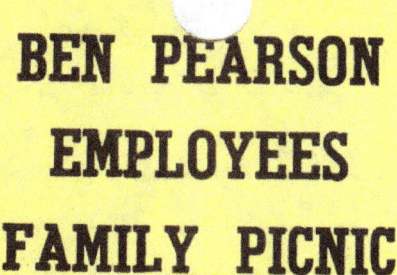

BEN PEARSON EMPLOYEES FAMILY PICNIC

August 26, 1961
3 P.M.
OAKLAND PARK

Bucky Maynard
NAME
8-26-61

J. B. "BUCKY" MAYNARD
MARKETING DIRECTOR

COLONIAL
LIFE & ACCIDENT INSURANCE COMPANY

2611 W. 34th
2–D
Pine Bluff, Ark. 71603
Tel.: 536–3222

Joe B. Maynard, Jr.

ASSISTANT MANAGER
NEWCOMER GREETING SERVICE

P.O. BOX 895
PINE BLUFF, ARK.
JE 4-6322

EDEN PARK COUNTRY CLUB
PINE BLUFF, ARKANSAS

JUNIOR MEMBER
1966 - 1967

Bucky Maynard

Signature *Bucky Maynard*

SATURDAY AFTERNOON DUPLICATE CLUB

JOE B. MAYNARD
DIRECTOR
TELEPHONE DAY 534-6322
NIGHT 535-0884

EDEN PARK COUNTRY CLUB
P.O. BOX 7858
PINE BLUFF, ARK.

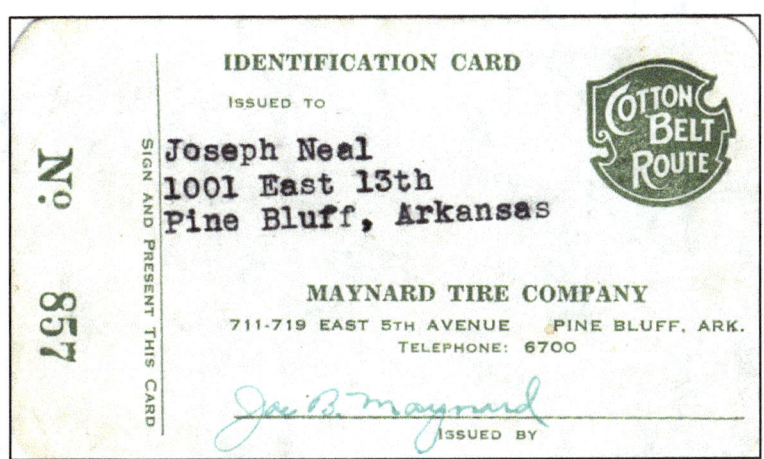

Newcomer Greetings Service Staff

Posted December 27, 2015 - 6:00am   Updated January 21, 2016 - 4:42pm

Joseph 'Bucky' Maynard to retire after 36 years

By Ocie Brown

CORRECTION: Note, the article below has not been changed, but the following correction has been issued: Joseph "Bucky" Maynard is a life-long Pine Bluff resident. An article in the Dec. 27 edition was incorrect as to when he moved to Pine Bluff. The newcomer welcome service he operates welcomes new residents to the area and represents all businesses, new and old. The article was incorrect about who the service deals with. Maynard started his insurance agency in 1979. The article was incorrect as to when he started his business.

* * *

After 36 years, Joseph 'Bucky' Maynard will be retiring at the end of the year.

Maynard is the owner of the J. B. Bucky Maynard Insurance company in Pine, Bluff, and since 1992 the company has served many residents and businesses around the city.

Before opening his insurance company, Maynard worked from home with Colonial Life. He was number one in sales in the United States in 1979, and from there he decided that this was the right business for him.

Prior to landing in Pine Bluff, Maynard worked as a Farmers Insurance Co. agent from 1982-1990. After having

http://pbcommercial.com/news/local/joseph-bucky-maynard-retire-after-36-years

Joseph 'Bucky' Maynard to retire after 36 years | Pine Bluff C

success as an agent for that company, Maynard became an independent agent in 1990, and he moved to Pine Bluff in 1992.

"We're independent, and we've insured a lot of different people and products throughout the city," Bucky said.

J. B. Bucky Maynard Insurance has worked with many different types of policies such as auto insurance, home owner's insurance, commercial, and claims.

Since starting his business, Maynard said he has encouraged his clients to be legal by helping them keep up with their liability insurance.

He said one of his proudest accomplishments during his career is that he's gotten a lot of people free medical insurance without receiving any commission.

"We understand that some people have low incomes and we've been able to keep rates low as possible in order to properly assist those clients," he said.

Over the years, Maynard has won numerous awards for sales and good performance.

Bucky Maynard said his son Lee Maynard will take over running the insurance company.

"He's been great to work for and he's built a successful business," Lee Maynard said. "I'm truly thankful that he's handing over the business to me and I plan to continue to run it how he has for years."

Now that Bucky Maynard will be retiring, he said he plans to continue to work with the newcomer welcome service that his dad started in 1952.

The newcomer welcome service contacts new local businesses that decide to move into town, in hopes of helping them advertise and gain exposure. Also after retiring, Bucky Maynard plans to finish and publish a book that he's started on.

"I want to thank the people in the Pine Bluff, White Hall and surrounding areas for their business and support," he said.

apx. 2011

Riverside

1-1-01

My Life and Thoughts

My Sci-Fi Short Story

It was a calm night at the Palomar Observatory in California. I had the often-mundane job of checking out the Hubbell telescope images from space. This particular evening we were tracking the famous Haley's Comet since it was coming closest to earth in 76 years. As I was watching the highly detailed images of the comet through the lenses of Hubbell for the first time ever I saw what appeared to be a chunk breaking off near the far almost unobservable side of the comet. The chunk almost seemed to make a zigzag motion before it disappeared behind the far side of Haley's, I instantly hit the replay button for the last several seconds to see if this was just my imagination. I played it back in slow motion and to my amazement the zigzag motion was even more noticeable.

I called my superior and got her home telephone answering machine. I get so tired of listening to a long message saying we are either away from the phone or not at home; etc. etc. and please leave a message after the beep. Why waste all this recorder time, we know the person is not going to answer and we know to leave a message after the beep. Why not just say leave a message and save all this time; Anyway I told her to call me that there was something peculiar on the picture of Haley's, I decided in the meantime to hook up a spectrometer to the Hubble images of Haley'. This machine shows different light waves coming from the comet. As I was watching these I noticed one colored wave that I recognized as only emanating from our new space plane developed last year.

This means that the temperature had to be in the range of a manmade object and not near what the temperature of Haley's Comet would be. I went back to the computer keyboard and typed a command to the telescope to focus in at a greater magnification. After this change took place it became clear that the zigzag object was what was generating the light spectrum with the manmade type temperature.

About this time the phone rang. Hello I said: "Doug this is Cheryl 0 – I got your message – what's going on she quizzed, after I told her, she said I'll be right down.

While I was waiting on Cheryl whom is really Dr. Cheryl Langston – Astrophysics, I noticed a second "chunk" that took a zigzag motion.

Later as we both watched the replay tapes in amazement we knew we had stumbled onto something unusual. *We had contacted the President's contact person and found that the President was ordering the new Space plane to fly in for a closer look. The space plane could fly to the far side of the comet to get a closer look at the mysterious objects that did not seem to be a part of Haley's original composition.*

We could monitor the audio part of the Space plane transmission – it was going like this. "Houston, we are now on the far side of the comet. It appears that there are several UFO's coming and going through an opening on this side of Haley's, for some reason they are becoming invisible to us as they get over 1 or 2 miles from the comet. I noticed on our UPS WIRE that Australia was suffering a giant earthquake over most of the whole continent. It had been going on for over 30 minutes. The UPS wire was going wild. A hunter in the outback reported seeing a strange purple laser type light earlier coming from the "sky" and focusing on an active volcano and that this beam seemed to seal off the volcano and cause it to stop spewing and smoking.

The President's spokesman called to say they were sending a scientist over who had some theories on what was happening. In the meantime strange things were happening all over the world –

a typhoon in the Pacific that was the biggest ever recorded, Northern lights appearing in the south; and the unending earthquake in Australia. As we were continuing our job of tracking Haley's the special scientist showed up and introduced himself as Dr. Fisher and that he had been studying this comet for quiet sometime. He said he had been under orders until now to keep it secret but that he had suspected for some time that aliens could be hiding in Haley's comet.

This was too much for me. I decided to take off a few hours for some much needed rest and to go home to my wife and kids so I would not be having a divorce anytime soon. When I got back about 6 hours later our observatory was abuzz with officials checking out our monitoring equipment. Dr. Fisher theorized that this had pretty well proven his research that he said went something like this.

From his studies and especially after today's weird phenomena that he thinks Haley's comet is not really the real Haley's but a replica of it. He theorized that it is really an alien spaceship whose mission is to cause us so much havoc that they can take us over. He said that the aliens and their ships were based on the silicon element instead of carbon as our bodies are based on. I remember from my high school chemistry teacher – the same one who told me that "fire is a liquid" – that elements with the same number of electrons in the outer shell react similarly.

Carbon has 4 electrons in the outer shell and so does silicon – the difference is that silicon has one extra of electron shells but still four in its outer shell. This means these 2 elements share a lot of the same properties. The difference is that glass is made of silicon and glass is transparent. So Dr. Fisher theorized that these beings were similar to us but that they appeared transparent or invisible to us.

This is why the UFO's disappeared a mile or 2 out.

As strange things were continuing to devastate the world Dr. Fisher said he had come up with a plan that might work. If his electromagnetic atomizer would work it would know the alien ships and even the aliens back to the carbon based structure and thus make them visible so we could better deal with them.

His plan was for our new space plane to get close enough to place some strategic laser beams on the fake comet to convert its contents over to the carbon type structure. As this was being carried out we asked what difference this would make – so what if we could see the alien ships, then what. The genius IQ Dr. Fisher said that if we could see them then the space plane could shoot them down with a new 'nuclear laser beam' that has only been in the experimental stages.

Well Dr. Fisher's first theory was working we were getting reports from all over the world that alien ships were mysteriously appearing out of nowhere all over the place.

After the space plane knocked out the one hovering over Australia the earthquake stopped 28 hours later. When asked why they did not fight back Dr. Fisher stated that even though the alien civilization was advanced on the molecular and chemical areas they were not a as far as we are into nuclear energy. i.e., they could become invisible but could not outfight us. As we were systematically destroying the alien ships, the fake Haley's comet's tail slowly started to smoke less and become shorter. Eventually the comet came to a dead stop and started to open up; as it did a huge oblong craft appeared out of it and headed toward earth. All our teletypes, fax machines, e-mail, computers, etc, said: PEACE do not be alarmed – you have won. If you will not destroy us we will show you the secret to invisibility, and hopefully the two worlds of ours can share a lot of information.

Well, a week later we were transferring all sorts of info. One of the best things we learned is that we could make our troops invisible and send them into battle here on earth undetected. This will make our wars easier to win here on earth but I wonder what the future will bring. What if we could make our whole planet invisible such as to hide from future invaders. Well for now I will contemplate my litter of 5 lab puppies and amuse myself as to how chaotic it would be if one of them turned out to be invisible.

By Bucky Maynard Nov.23, 2001 ...

www.ingramcontent.com/pod-product-compliance
Lightning Source LLC
Chambersburg PA
CBHW072107290426
44110CB00014B/1861